# TABLE OF CONTEN

OTM2159   ISBN: 9781770783515
© On The Mark Press

# Teacher Assessment Rubric

Student's Name: _____     Date: _____

| Success Criteria | Level 1 | Level 2 | Level 3 | Level 4 |
|---|---|---|---|---|
| **Knowledge and Understanding Content** | | | | |
| Demonstrate an understanding of the concepts, ideas, terminology definitions, procedures and the safe use of equipment and materials | Demonstrates limited knowledge and understanding of the content | Demonstrates some knowledge and understanding of the content | Demonstrates considerable knowledge and understanding of the content | Demonstrates thorough knowledge and understanding of the content |
| **Thinking Skills and Investigation Process** | | | | |
| Develop hypothesis, formulate questions, select strategies, plan an investigation | Uses planning and critical thinking skills with limited effectiveness | Uses planning and critical thinking skills with some effectiveness | Uses planning and critical thinking skills with considerable effectiveness | Uses planning and critical thinking skills with a high degree of effectiveness |
| Gather and record data, and make observations, using safety equipment | Uses investigative processing skills with limited effectiveness | Uses investigative processing skills with some effectiveness | Uses investigative processing skills with considerable effectiveness | Uses investigative processing skills with a high degree of effectiveness |
| **Communication** | | | | |
| Organize and communicate ideas and information in oral, visual, and/or written forms | Organizes and communicates ideas and information with limited effectiveness | Organizes and communicates ideas and information with some effectiveness | Organizes and communicates ideas and information with considerable effectiveness | Organizes and communicates ideas and information with a high degree of effectiveness |
| Use science and technology vocabulary in the communication of ideas and information | Uses vocabulary and terminology with limited effectiveness | Uses vocabulary and terminology with some effectiveness | Uses vocabulary and terminology with considerable effectiveness | Uses vocabulary and terminology with a high degree of effectiveness |
| **Application of Knowledge and Skills to Society and Environment** | | | | |
| Apply knowledge and skills to make connections between science and technology to society and the environment | Makes connections with limited effectiveness | Makes connections with some effectiveness | Makes connections with considerable effectiveness | Makes connections with a high degree of effectiveness |
| Propose action plans to address problems relating to science and technology, society, and environment | Proposes action plans with limited effectiveness | Proposes action plans with some effectiveness | Proposes action plans with considerable effectiveness | Proposes action plans with a high degree of effectiveness |

OTM2159   ISBN: 9781770783515
© On The Mark Press

# Student Self-Assessment Rubric

Name: _____ Date: _____

Put a check mark ✔ in the box that best describes you.

| Expectations | Always | Almost Always | Sometimes | Seldom |
|---|---|---|---|---|
| I listened to instructions. | | | | |
| I was focused and stayed on task. | | | | |
| I worked safely. | | | | |
| My answers show thought, planning, and good effort. | | | | |
| I reported the results of my experiment. | | | | |
| I discussed the results of my experiment. | | | | |
| I used science and technology vocabulary in my communication. | | | | |
| I connected the material to my own life and the real world. | | | | |
| I know what I need to improve. | | | | |

1. I liked _____

_____

2. I learned _____

_____

3. I want to learn more about _____

_____

# INTRODUCTION

The activities in this book have two intentions: to teach concepts related to earth and space science, and to provide students the opportunity to apply necessary skills needed for mastery of science and technology curriculum objectives.

Throughout the experiments, the scientific method is used. The scientific method is an investigative process which follows five steps to guide students to discover if evidence supports a hypothesis.

1. **Consider a question to investigate.**
   For each experiment, a question is provided for students to consider. For example, "What effect do you think salt has on the density of water?

2. **Predict what you think will happen.**
   A hypothesis is an educated guess about the answer to the question being investigated. For example, "I believe that salted water is denser and can support the mass of certain objects". A group discussion is ideal at this point.

3. **Create a plan or procedure to investigate the hypothesis.**
   The plan will include a list of materials and a list of steps to follow. It forms the "experiment".

4. **Record all the observations of the investigation.**
   Results may be recorded in written, table, or picture form.

5. **Draw a conclusion.**
   Do the results support the hypothesis? Encourage students to share their conclusions with their classmates, or in a large group discussion format.

The experiments in this book fall under ten topics that relate to one complete aspect of earth and space science: **Fresh and Salt Water Systems on Earth**. In each section you will find teacher notes designed to provide you guidance with the learning intention, the success criteria, materials needed, a lesson outline, as well as provide some insight on what results to expect when the experiments are conducted. Suggestions for differentiation are also included so that all students can be successful in the learning environment.

## ASSESSMENT AND EVALUATION:

Students can complete the Student Self-Assessment Rubric in order to determine their own strengths and areas for improvement. Assessment can be determined by observation of student participation in the investigation process. The classroom teacher can refer to the Teacher Assessment Rubric and complete it for each student to determine if the success criteria outlined in the lesson plan has been achieved. Determining an overall level of success for evaluation purposes can be done by viewing each student's rubric to see what level of achievement predominantly appears throughout the rubric.

OTM2159   ISBN: 9781770783515
© On The Mark Press

# WATER IN OUR WORLD

## LEARNING INTENTION:
Students will learn about the water cycle, forms of water, and world distribution of water.

## SUCCESS CRITERIA:
- describe how the water cycle creates precipitation in our natural environment
- recreate the water cycle
- identify water in different forms and conditions
- determine percentages of our world water distribution
- research facts about salt and fresh water

## MATERIALS NEEDED:
- a copy of "The Water Cycle" Worksheet 1 for each student
- a copy of "Exploring the Water Cycle" Worksheet 2 and 3 for each student
- a copy of "Forms of Water" Worksheet 4 for each student
- a copy of "The Water on our Earth" Worksheet 5 for each student
- a copy of "Getting the Facts!" Worksheet 6 for each student
- a clear glass jar, a foil pan, about 6 ice cubes (a set for each group of students)
- a large jug of water, a few desk lamps, a large funnel

## PROCEDURE:
*This lesson can be done as one long lesson, or be divided into three or four shorter lessons.*
1. Give students Worksheet 1. Read through the information about nature's water cycle. Discuss the concepts of 'condensation', 'precipitation', and 'evaporation' with students to ensure their understanding of how each is created and part of the cycle.

2. Explain to students that they will recreate the water cycle. Divide students into groups and give them Worksheets 2 and 3, and the materials to conduct the examination. Read through the materials needed and what to do sections with students to ensure their understanding of the task. Students will record their observations and conclusion on Worksheet 3.

3. Give students Worksheet 4. They will engage in a 'think-pair-share' activity to discuss and record forms and conditions in which water exists in the environment. Sample responses:

| | |
|---|---|
| **Solid** | • ice, snow, hail, sleet, frost<br>• present in glaciers, polar ice caps |
| **Gas** | • fog, water vapor, clouds<br>• present as a gas in the atmosphere |
| **Liquid** | • rain, dew<br>• present in oceans, lakes, rivers, ponds, streams, in ground water |

4. Give students Worksheet 5. With access to the internet, they will research the percentages of salt water, fresh water, solid water, and gaseous water on our Earth. The data is to be presented in a pie chart.

5. Give students Worksheet 6. With access to the internet, they will research interesting facts about salt water and fresh water on our Earth. The facts are to be organized in the T-chart.

## DIFFERENTIATION:
Slower learners may benefit by working as one small group with teacher support to complete Worksheet 4 together on chart paper. An additional accommodation could be to discuss the 'Challenge Question' on Worksheet 3, as a small group, in place of individual written responses.

**For enrichment**, faster learners could label a world map, indicating world water distribution. They should locate oceans, bodies of fresh water, polar ice caps, presence of ground water, etc.

OTM2159   ISBN: 9781770783515
© On The Mark Press

# The Water Cycle

Water exists in three different forms. It can be a solid, liquid, or a gas. Let's learn exactly how these forms of water cycle through the environment.

**Condensation** happens when water in a gas form meets cooler air. It will form a cloud in the sky where this gas changes into droplets of water.

These droplets of water, called precipitation, fall back down to Earth. **Precipitation** can be in the form of rain, snow, sleet, or hail. It wets the ground, and fills up rivers, lakes, oceans, and streams.

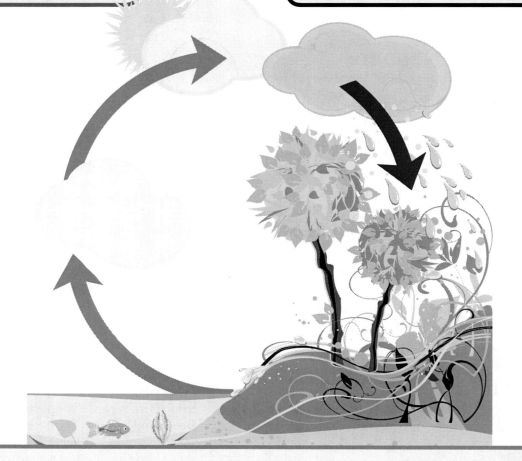

As the sun heats the ground, oceans, rivers, lakes, and streams, the water evaporates. **Evaporation** is water in its gas form rising again to condense as a cloud.

 OTM2159   ISBN: 9781770783515
© On The Mark Press

# Exploring the Water Cycle

Create your own water cycle. Try this!

## You'll need:

- a clear glass jar, half filled with water
- a foil pan
- about 6 ice cubes
- a desk lamp

## What to do:

1. Place the foil pan on top of the glass jar that is half filled with water.

2. Put the ice cubes on the foil pan.

3. Shine the light from the lamp directly at the ice cubes.

4. Make observations and record them on Worksheet 3.

5. Make conclusions and record them on Worksheet 3.

OTM2159   ISBN: 9781770783515
© On The Mark Press

## Let's Observe

Describe what you see sticking to the bottom of the aluminum pan inside the glass.

_____

_____

_____

If the water in the bottom of the jar represents a lake, then:
What do you think the lamp represents?

_____

_____

What do you think the aluminum pan and the ice cubes represent?

_____

_____

## Let's Conclude

Use your observations to explain what is happening inside the jar.

_____

_____

_____

_____

Challenge question:

Do you think the water in the jar will ever get used up? Explain your thinking.

_____

_____

_____

_____

_____

OTM2159   ISBN: 9781770783515
© On The Mark Press

Name:

# Forms of Water

You have learned that water exists in three different states. These states are as a solid, as a liquid, and as a gas.

**With a partner**, do some thinking and sharing of ideas about the forms and conditions in which these three states of water exist in our environment. Record your ideas below.

**Water as a solid**



**Water as a liquid**



**Water as a gas**

# The Water on our Earth

Do some research in order to create a pie chart that illustrates the water distribution on our Earth. Your graph should indicate the percentages of:

- salt water
- fresh water
- solid water
- gaseous water

**Title:** _____

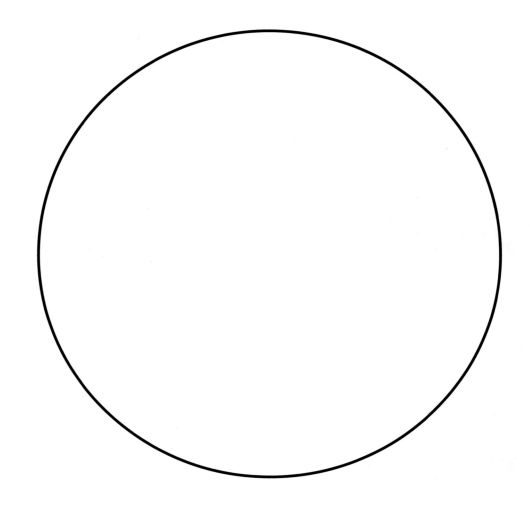

Explain the importance of fresh water conservation. Justify your response.

_____

_____

_____

_____

OTM2159   ISBN: 9781770783515
© On The Mark Press

Name:

# Getting the Facts!

Continue your research about water. Research and record some interesting facts about salt water and fresh water in our world. Use the T-chart below to organize your findings.

| Salt Water | Fresh Water |
| --- | --- |
| | |

# SALT AND FRESH WATER

## LEARNING INTENTION:
Students will learn about some of the differences between salt water and fresh water.

## SUCCESS CRITERIA:
- compare the density of fresh water to the density of salt water
- determine the freezing points of fresh water and salt water
- determine the boiling points of fresh water and salt water
- use charts and graphs to display observations
- make conclusions about the differences of fresh water and salt water
- make connections to environment of our world

## MATERIALS NEEDED:
- a copy of "A Comparison of Density" Worksheet 1, 2, and 3 for each student
- a copy of "Investigating a Freezing Point" Worksheet 4, 5, and 6 for each student
- a copy of "Investigating a Boiling Point" Worksheet 7, 8, and 9 for each student
- 2 glass jars, 2 plastic containers, 2 hot plates, 2 centigrade thermometers, a tablespoon, a ruler, a marker, 2 drinking straws, 6 cups of water, 2 flasks with rubber stoppers that can hold thermometers (a set for each group)
- table salt, modeling clay, labels, timers, access to a freezer
- atlases, access to the internet, clipboards, pencils

## PROCEDURE:
*This lesson can be done as one long lesson, or divided into three shorter lessons.
1. Give students Worksheet 1. Read through the question, materials needed, and what to do section to ensure students' understanding. Divide students into small groups, and give them the materials and Worksheets 2 and 3 to complete the investigation on the density of the waters.

2. Give students Worksheet 4. Read through the question, materials needed, and what to do section to ensure students' understanding. Divide students into small groups, give them the materials and Worksheets 5 and 6 to complete as they conduct the investigation on the freezing points of fresh and salt water. An option is to come together as a large group to discuss their findings on the 'Let's Connect It!' section on Worksheet 6. **(Icy roadways in cold winter climates are treated with sand rather than salt. Due to salt water's freezing temperature being about -20°C, it would be ineffective in climates colder than that.)**

3. Give students Worksheet 7. Read through the question, materials needed, and what to do section to ensure students' understanding. Divide students into small groups, give them the materials and Worksheets 8 and 9 to complete as they conduct the investigation on the boiling points of fresh and salt water. *An option is to organize this investigation as a teacher led experiment with the large group observing. Time and temperature readings could be recorded on chart paper, then students could complete Worksheet 9 individually using the data that was collected as a group.*

4. Direct students to add facts that they have learned from the three experiments to the T-chart on 'Getting the Facts!' Worksheet 6 from the previous lesson.

## DIFFERENTIATION:
Slower learners may benefit by working with a strong peer to collect data in each experiment. Interpreting the data in the 'Let's Connect It!' sections, could be done as a small group with teacher support to ensure that students understand the concept.

**For enrichment,** faster learners could continue to boil the waters until full evaporation occurs, then examine the differences in residue.

OTM2159  ISBN: 9781770783515
© On The Mark Press

# A Comparison of Density

**Density** is a factor in whether something will sink or float. In this experiment, you will be investigating the density of salt water and fresh water. You will create your own hydrometer, which is a tool that measures and compares the density of one liquid to another. Let's get started!

**Question:** What affect do you think salt has on the density of water?

**You'll need:**

- 2 glass jars
- 2 cups of water
- a spoon
- table salt

- a ruler
- 2 small pieces of modeling clay
- a marker
- 2 drinking straws

**What to do:**

1. Make a prediction to the question. Record it on Worksheet 2.

2. Attach a small ball of modeling clay to one end of each of the straws.

3. Using the ruler and marker, measure and mark a scale along each of the straws, using increments of 5 mm. You will have created 2 hydrometers.

4. Pour 1 cup of water into a glass, then place one of the hydrometers into the glass of water, with modeling clay towards the bottom.

5. Observe the point where the surface of the fresh water measures on the hydrometer. Record it on Worksheet 2. Draw and label a diagram to illustrate your observations.

6. Pour 1 cup of water into another glass. Add the salt, one spoonful at a time, and stir until it becomes saturated (this is means no more salt can be dissolved).

7. Place the other hydrometer into the glass of salted water, with modeling clay towards the bottom.

8. Repeat step 5.

9. On Worksheet 3, make a conclusion and connections about what you have observed.

Name:

## Let's Predict

What effect do you think salt has on the density of water?

_____

_____

_____

## Let's Observe

Draw and label a diagram that illustrates what you have observed when the hydrometer was placed in **fresh water**.

Provide a written description of your observations:

_____

_____

_____

_____

_____

_____

_____

Draw and label a diagram that illustrates what you have observed when the hydrometer was placed in **salt water**.

Provide a written description of your observations:

_____

_____

_____

_____

_____

_____

OTM2159   ISBN: 9781770783515
© On The Mark Press

## Let's Conclude

Write a concluding statement about the density of fresh water compared to the density of salt water. Use the data you collected and your observations to support your answer.

_____

_____

_____

_____

## Let's Connect It!

A cargo ship loads up in a port on Lake Ontario. It will be traveling to Europe to deliver its goods. On the journey the ship will sail on Lake Ontario, then travel down the St. Lawrence River, and continue out across the Atlantic Ocean to Europe.

Use an atlas to help you map out the ship's route.

Will there be differences in the ship's buoyancy as it makes its journey to Europe? Explain your thinking.

_____

_____

_____

# Investigating a Freezing Point

**Question:** Is there a difference between the freezing points of fresh water and salt water?

**You'll need:**

- 2 plastic containers
- 2 centigrade thermometers
- a spoon
- table salt
- a timer

- 2 cups of water
- 2 labels
- access to a freezer
- a marker

**What to do:**

1. Make a prediction to the answer of the question. Record it on Worksheet 5.

2. Pour 1 cup of water into one of the plastic containers. Label this **"freshwater"**.

3. Pour 1 cup of water into the other plastic container, add the salt and stir until it is saturated (this means no more salt can be dissolved). Label this **"salt water"**.

4. Place a thermometer into each container. On Worksheet 5, record the temperature of the waters.

5. Place the containers of water in the freezer, and set the timer for 30 minutes.

6. After 30 minutes has passed, take the temperature of the waters again. Record the data on Worksheet 5.

7. Repeat steps 5 and 6, every 30 minutes, until about 6 hours have passed.

8. Make a conclusion about the freezing points of fresh water and salt water from the data you collected and the observations you have made. Record your conclusion on Worksheet 6.

9. Using your conclusion, make connections to the environment of our world. Record them on Worksheet 6.

OTM2159   ISBN: 9781770783515
© On The Mark Press

Name:

## Let's Predict

Is there a difference between the freezing points of fresh water and salt water? _____

_____

## Let's Observe

Make observations and record the temperatures of the water in the chart below.

| Temperature of Fresh Water | Observations of Fresh Water Consistency | Temperature of Salt Water | Observations of Salt Water Consistency |
|---|---|---|---|
| | | | |
| | | | |
| | | | |
| | | | |
| | | | |
| | | | |
| | | | |
| | | | |
| | | | |
| | | | |
| | | | |
| | | | |
| | | | |

OTM2159   ISBN: 9781770783515
© On The Mark Press

Name: _____

## Let's Conclude

Write a concluding statement about the differences in the freezing point of fresh water compared to salt water. Use your data and your observations to support your answer.

_____

_____

_____

_____

## Let's Connect It!

1) Place a thermometer in your freezer. Wait 15 minutes, and read the temperature.

   The temperature inside the freezer is _____ .

2) Use the internet to research the freezing point of salt water.

   The freezing point of salt water is _____ .

3) Using the information you have gathered, complete the sentences below.

   The temperature of my freezer is _____ , so the salt

   water solution did not completely _____ .

   If the temperature of my freezer was _____ , the salt

   water solution would have _____ .

4) How is this information useful to crews who maintain roadways in places that have cold winter climates?

   _____

   _____

   _____

   _____

   _____

OTM2159   ISBN: 9781770783515
© On The Mark Press

# Investigating a Boiling Point

**Question:** Is there a difference between the boiling points of fresh water and salt water?

**You'll need:**

- 2 hot plates
- 2 centigrade thermometers
- table salt
- 2 cups of water

- 2 timers
- 2 flasks with rubber stoppers that can hold thermometers
- a spoon
- a partner

**What to do:**

1. Make a prediction to the answer of the question. Record it on Worksheet 8.

2. Pour 1 cup of water into one of the flasks. Put the rubber stopper on and place a thermometer inside of the flask, through the rubber stopper.

3. Pour 1 cup of water into the other flask, add the salt and stir until it is saturated (this means no more salt can be dissolved). Put the rubber stopper on and place a thermometer inside of the flask, through the rubber stopper.

4. On Worksheet 8, record the temperature of the waters.

5. Place each of the flasks on a hot plate. One partner will record the temperature of the water, while the other partner records the temperature of the salted water, every 30 seconds. (Each partner will need his/her own timer).

6. When the waters begin to bubble, the boiling point is reached, and the temperature will level off. On Worksheet 8, record the amount of time that each of the waters took to reach their boiling point. (*Continue recording the temperature of the boiling waters for 2 minutes after each has reached its boiling point).

7. On Worksheet 9, create a line graph that displays the data that you collected.

8. Make a conclusion about the boiling points of fresh water and salt water from the data you collected. Record your conclusion on Worksheet 9.

## Let's Predict

Is there a difference between the boiling points of fresh water and salt water? _____

_____

## Let's Observe

In the chart, record the time and temperatures of the waters as they come to a boil.

| Time (fresh water) | Temperature of Fresh Water | Time (salt water) | Temperature of Salt Water |
|---|---|---|---|
| | | | |
| | | | |
| | | | |
| | | | |
| | | | |
| | | | |
| | | | |
| | | | |
| | | | |
| | | | |
| | | | |
| | | | |

OTM2159   ISBN: 9781770783515
© On The Mark Press

## Let's Conclude

Create a line graph to display the data that you collected in the chart on Worksheet 8.

_____

1) The boiling point of fresh water is _____ .

2) The boiling point of salt water is _____ .

3) Write a concluding statement about the differences in the boiling point of fresh water compared to salt water. Use your data to support your answer.

_____

_____

_____

_____

# WATERSHEDS

## LEARNING INTENTION:

Students will learn about continental divides, the directional flow of the water, and how ocean currents are connected on a global scale.

## SUCCESS CRITERIA:

- determine the location of the continental divides in North America
- identify the watersheds separated by these continental divides
- determine how a local body of water can ultimately flow into an ocean of the world
- describe and recreate the ocean thermo-haline system
- make connections to the health of the oceans and marine life in them

## MATERIALS NEEDED:

- a copy of "Which Way Does the Water Flow?" Worksheet 1 and 2 for each student
- a copy of "Go With the Flow!" Worksheet 3 and 4 for each student
- a copy of "The Thermo-haline System" Worksheet 5 for each student
- a copy of "Recreating the Global Conveyer Belt" Worksheet 6, 7, and 8 for each student
- atlases, large maps of your area, access to the internet
- red food coloring, a kettle, access to tap water
- 6 Styrofoam cups, a large clear rectangular tub, a spoon, about 6 blue ice cubes (for each group of students)
- clipboards, pencils

## PROCEDURE:

*This lesson can be done as one long lesson, or be divided into three shorter lessons.

1. Give students Worksheets 1 and 2. Read through with the students about continental divides. Students will access the internet or atlases to learn about the location of the continental divides in North America, then they will indicate them on the map on Worksheet 1. Using Worksheet 2, students will describe the watersheds created by each continental divide.

2. Give students Worksheets 3 and 4. Using large maps of the local area, atlases, and the internet, they will track the route that a pail of water would travel from where they live to its ultimate destination (one of the oceans). Read through the criteria with students to ensure their understanding of the task.

3. Give students Worksheet 5. Read through and discuss with students the information about the ocean thermo-haline system to ensure their understanding of the concept.

4. Divide students into small groups and give them Worksheets 6, 7, and 8, and the materials to conduct the investigation. Read through the question, materials needed, and what to do sections to ensure their understanding of the task. They will make a prediction, conduct the investigation, record observations, make a conclusion, and connect their learning to the environment and the ocean world.

## DIFFERENTIATION:

Slower learners may benefit by working as one small group with teacher support to complete Worksheet 3 and 4 together on chart paper. An additional accommodation could be to discuss the questions in the 'Let's Connect It!' section on Worksheet 9, as a small group, in place of individual written responses.

**For enrichment,** faster learners could use the internet to research the role of an oceanographer.

- what can they specialize in?
- what are some typical working tasks of an oceanographer?

OTM2159  ISBN: 9781770783515
© On The Mark Press

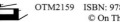

# Which Way Does the Water Flow?

In North America, there is a long mountainous ridge and an ice field of many glaciers that divides the continent. This divide is called the Great Continental Divide. It spans from northern Canada, through the United States, and into Mexico. The waters on either side of the Great Continental Divide flow into the Pacific Ocean, the Atlantic Ocean, or the Arctic Ocean. While there is only one Great Continental Divide in North America, this continent borders more than two bodies of water, so it has more than one continental divide.

North America has five continental divides. Do some research to find out where they are located, and then label them on the map of North America.

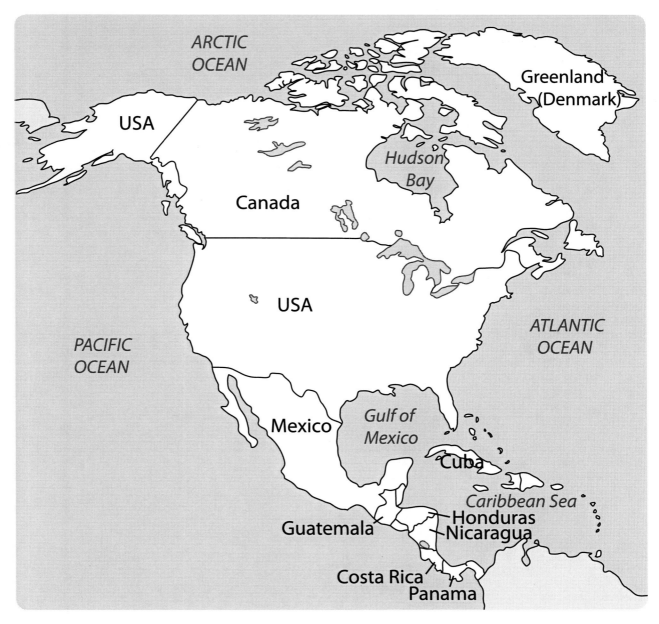

For each of the continental divides that you labeled on the map on Worksheet 1, identify the watersheds that they separate.

1) The _____ Divide.

_____

_____

_____

2) The _____ Divide.

_____

_____

_____

3) The _____ Divide.

_____

_____

_____

4) The _____ Divide.

_____

_____

_____

5) The _____ Divide.

_____

_____

_____

OTM2159   ISBN: 9781770783515
© On The Mark Press

# Go With the Flow!

Access a map of your area that indicates local bodies of water, and depicts their connections to larger bodies of water. Your task is to describe the route a pail of water would flow from your community to its ultimate destination.

**Details to include:**

- detailed illustration of its route
- names of bodies of water it will flow through
- directions of their flow
- the continental divide(s) that separate their watersheds
- approximate number of kilometres it will travel to reach its ultimate destination

Detailed illustration of its route:

The water from the pail will begin at

_____

_____

_____

_____

It will continue to travel

_____

_____

_____

_____

_____

_____

_____

_____

_____

_____

_____

_____

_____

_____

_____

_____

_____

_____

_____

_____

reaching its ultimate destination, which is _____ ,

traveling a total of _____ kilometres.

OTM2159   ISBN: 9781770783515
© On The Mark Press

# The Thermo-haline System

The water in the oceans is always moving, creating currents. Did you know that these ocean currents are connected on a global scale?

To understand how this happens, it is important to know that cold salt water in the ocean is denser than its warmer water, so it sinks. In relation to the Earth, the cold salt water at the North Pole and the South Pole sinks and flows toward the equator. The water at the equator is warm and it gets pushed upward even more by the cold water that is coming underneath it.

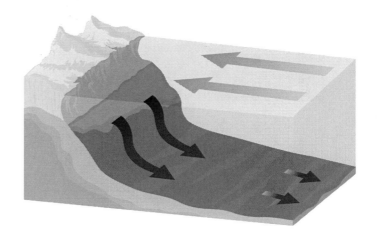

As the warm water travels along the surface of the ocean, it moves further away from the equator, causing it to become cooler. As it goes from cool to cold, it sinks, which completes a cycle, like a conveyer belt. This cycle is known as the thermo-haline system.

The convection currents that are continuously circulating in the oceans help to redistribute nutrients and oxygen.

The ocean waves mix water with air. As surface water cools and sinks down, it mixes with bottom waters which push up when they are warmed.

This creates an overturn of nutrients and oxygen for the living matter in the oceans.

OTM2159   ISBN: 9781770783515
© On The Mark Press

# Recreating the Global Conveyer Belt

You have learned about directional flow of water and how it continuously moves in a circular pattern in the oceans of our world, like a conveyer belt. Now, let's take a closer look at how this works!

**Question:** How will cold water react to contact with hot water?

**You'll need:**

- 6 Styrofoam cups
- a large, clear, rectangular tub
- red food coloring
- access to tap water
- a kettle
- a spoon
- a tray of blue ice cubes (add blue food coloring to the water before freezing)

**What to do:**

1. Make a prediction to the question. Record it on Worksheet 7.

2. Fill the clear tub with room temperature water (about ¾ full).

3. Turn 4 Styrofoam cups upside down on a table. Next, place the tub of water on top of them, so that each cup is supporting a corner of the tub. Allow the water to settle.

4. Put some of the blue ice cubes into the water, at one end of the tub. Observe what happens. Record your observations on Worksheet 7.

5. Pour very hot water into 2 Styrofoam cups. Carefully place one of these cups underneath the tub, at the opposite end from the ice cubes.

6. In the other cup filled with hot water, add some red food coloring and stir.

7. Slowly pour the red, hot water into the tub of water, at the opposite end from the blue ice cubes. Observe what happens. Record your observations on Worksheet 7.

8. On Worksheet 8, make a conclusion and connections about what you have observed.

OTM2159   ISBN: 9781770783515
© On The Mark Press

## Let's Predict

How will cold water react to contact with hot water?

_____

_____

_____

## Let's Observe

Draw and label a diagram that illustrates what you have observed when the blue ice cubes were placed into the tub of water.

Provide a written description of your observations:

_____

_____

_____

_____

_____

_____

_____

Draw and label a diagram that illustrates what you have observed when the red, hot water was placed into the water at the other end of the tub.

Provide a written description of your observations:

_____

_____

_____

_____

_____

_____

_____

OTM2159   ISBN: 9781770783515
© On The Mark Press

Name: _____

## Let's Conclude

Write a concluding statement about how cold water reacts to contact with hot water. Use your observations to support your answer.

_____

_____

_____

_____

## Let's Connect It!

Why is the thermo-haline system important for the health of the oceans and the marine life in them? Explain your thinking.

_____

_____

_____

_____

_____

_____

_____

How could rising temperatures caused by global warming affect this global ocean conveyor belt? Explain your thinking.

_____

_____

_____

_____

_____

_____

_____

OTM2159   ISBN: 9781770783515
© On The Mark Press

# EFFECTS ON THE WATER

## LEARNING INTENTION:

Students will learn about the effects of wind currents in the oceans on climate and the effects of global warming on glaciers and ocean levels.

## SUCCESS CRITERIA:

- identify the five main gyres in the oceans of the world
- describe the effects of islands and banks on ocean currents
- research a gyre to determine the effects on climate associated with its currents
- describe and locate two types of glaciers in our world
- evaluate the impacts of melting ice packs and melting glaciers on sea levels

## MATERIALS NEEDED:

- a copy of "Wind in the Ocean" Worksheet 1, 2, 3, and 4 for each student
- a copy of "Ocean Currents and Climate" Worksheet 5, 6, and 7 for each student
- a copy of "Glaciers" Worksheet 8 for each student
- a copy of "Glaciers and Sea Level" Worksheet 9, 10, 11, and 12 for each student
- a glass Petrie dish, a clear rectangular tub, a few straws, a tall rock (a set for each group of students)
- food coloring, access to tap water, plasticine, table salt, a few tablespoons
- access to the internet
- Bristol board (once piece per group)
- a cutting board, a clear rectangular container (about 9" x 13'), a ruler, 2 trays of ice cubes (a set for each group of students)
- clipboards, pencils

## PROCEDURE:

*This lesson can be done as one long lesson, or be divided into four shorter lessons.

1. Give students Worksheet 1. Read through with students, the information on how the winds patterns of the world affect the ocean currents. Discuss the concepts of 'prevailing winds', 'northern and southern hemispheres', and 'gyres' with students to ensure their understanding of how they are connected to the direction of ocean currents.

2. Explain to students that they will examine the effects of islands and banks on ocean currents. Divide students into groups and give them Worksheets 2, 3, and 4, and the materials to conduct the examination. Read through the materials needed and what to do sections with students to ensure their understanding of the task. As they investigate, students will record their observations and conclusion on Worksheets 3 and 4.

3. Give students Worksheets 5, 6, and 7. With access to the internet, they will research one of the five main gyres in the oceans, to learn more about its effect on the climates of cities within its location.

4. Once the assignment on gyres is completed, conduct a jigsaw activity where students are assigned to small groups of 5, in which each student in the group has prepared information on a different gyre. Students will present their information to the other members in their group. *A follow-up option is to give the groups another copy of the world map to plot their gyres and cities on, as one final copy. They can glue it in the centre of a piece of Bristol board, and then glue all of their copies of Worksheet #6 around it. This would make an interesting bulletin board display.*

5. Give students Worksheet 8. Read through with students the information on the different types of glaciers and their locations. Discuss the concepts with students to ensure their understanding.

6. Explain to students that they will investigate the impact of global warming and the effects of melting glaciers on sea levels. Divide students into groups and give them Worksheets 9, 10, 11, 12, and the materials to conduct the investigation. Read through the question, materials needed, and what to do sections with students to ensure they understand the task. As they investigate, students will record their observations of water level in a chart on Worksheet 10, display their data in a line graph on Worksheet 11, and make conclusions and a connection on Worksheet 12.

## DIFFERENTIATION:

Slower learners may benefit by:

- completing only the diagrams on Worksheets 3 and 4, then discussing their concluding ideas and the challenge question in a small group, with teacher support

- working in a small group with teacher direction to research and compile information about a gyre on Worksheets 5, 6, and 7 (this would ensure that when they jigsaw into other groups, they have accurate information to share)

- working in a small group with teacher support to accurately display the data that they collected on Worksheet 10 in a line graph on Worksheet 11

**For enrichment**, faster learners could:

- use the information they gathered on Worksheets 3 and 4 "Go With the Flow" (from last lesson) to describe the effect a rise in water level in the oceans, due to global warming and glacial melt, would have on the water levels in bodies of water in their locality

- access the internet to research various alpine and continental glacial features, then illustrate them (e.g., cirques, arêtes, horns, hanging valleys, crevasses, moraines, eskers, outwash, fiords, icebergs, striations)

OTM2159  ISBN: 9781770783515
© On The Mark Press

# Wind in the Ocean

Prevailing winds of the Earth develop waves and push them in one predominant direction, which causes the formation of surface ocean currents. These prevailing winds travel in a clockwise motion in the northern hemisphere, and in a counter-clockwise motion in the southern hemisphere.

These currents travel in consistent, circular patterns, creating gyres, which are spiraling currents on the surface of the ocean. As the currents in the gyres circulate, an overturn of nutrients and oxygen is also created.

There are five main gyres in the world. They are the North Atlantic gyre, the South Atlantic gyre, the South Indian gyre, the North Pacific gyre, and the South Pacific gyre.

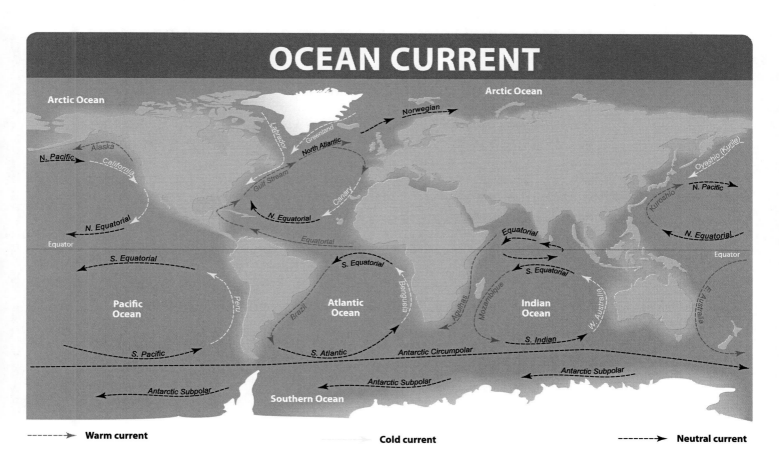

**Question:** What are the effects of islands and banks on currents in the ocean?

**You'll need:**

- a glass Petrie dish
- a clear rectangular tub
- food coloring
- access to tap water
- a few straws
- a tall rock (that will stand higher than water level in tub)

**What to do:**

1. Make a prediction to the question. Record it on worksheet 3.

2. Fill the clear tub with water (about ½ full). Place it on a table and allow the water to settle.

3. Place a few drops of food coloring at one end of the tub. Using a straw, gently blow across the water. Record your observations on worksheet 3.

4. Place a tall rock in the centre of the tub. This will represent an island.

5. Place a few drops of food coloring at one end of the tub. Using a straw, gently blow across the water. Observe what happens in the front and the back of the island. Record your observations on worksheet 3.

6. Remove the rock. Place the Petrie dish (face down) in its place, on the bottom of the tub. This will represent a bank.

7. Place a few drops of food coloring at one end of the tub. Using a straw, gently blow across the water. Record your observations on worksheet 4.

8. On worksheet 4, make a conclusion about what you have observed.

9. With a partner or small group, discuss the challenge question on worksheet 4. Record your response.

OTM2159  ISBN: 9781770783515
© On The Mark Press

## Let's Predict

What are the effects of islands and banks on currents in the ocean?

_____

_____

_____

## Let's Observe

Draw and label a diagram that illustrates what you observed when you created wind across the surface of the water in step 3.

Provide a written description of your observations:

_____

_____

_____

_____

_____

_____

_____

Draw and label a diagram that illustrates what you observed when you created wind across the surface of the water in step 5.

Provide a written description of your observations:

_____

_____

_____

_____

_____

_____

_____

Draw and label a diagram that illustrates what you observed when you created wind across the surface of the water in step 7.

Provide a written description of your observations:

_____

_____

_____

_____

_____

_____

_____

## Let's Conclude

Write a concluding statement about the effects of islands and banks on currents in the ocean. Use your observations to support your answer.

_____

_____

_____

_____

_____

_____

## Challenge question!

How do you think wind direction and ocean current affect inland weather patterns? _____

_____

_____

_____

_____

OTM2159   ISBN: 9781770783515
© On The Mark Press

# Ocean Currents and Climate

You have learned a lot about the circulation of ocean waters, and how wind affects the currents. Your task now is to find out more about **one** of the five main gyres in the oceans and how climates in its proximity are affected by it.

## Let's Research

**Gyre Name:** _____

**Location:** _____

**List the currents that make up this gyre:**

- _____
- _____
- _____
- _____
- _____
- _____
- _____
- _____
- _____

**Name 2 cities at different points on this gyre:**

- _____
- _____

OTM2159   ISBN: 9781770783515
© On The Mark Press

**City #1** _____ :

Average temperature: _____

Average precipitation: _____

**City #2** _____ :

Average temperature: _____

Average precipitation: _____

**Use the data you collected to explain the climate of City #1 because of the associated ocean currents in the gyre.**

_____

_____

_____

_____

_____

**Use the data you collected to explain the climate of City #2 because of the associated ocean currents in the gyre.**

_____

_____

_____

_____

_____

_____

OTM2159  ISBN: 9781770783515
© On The Mark Press

Name:

Use a black marker to label your gyre on the map. Next, pinpoint and label your cities.

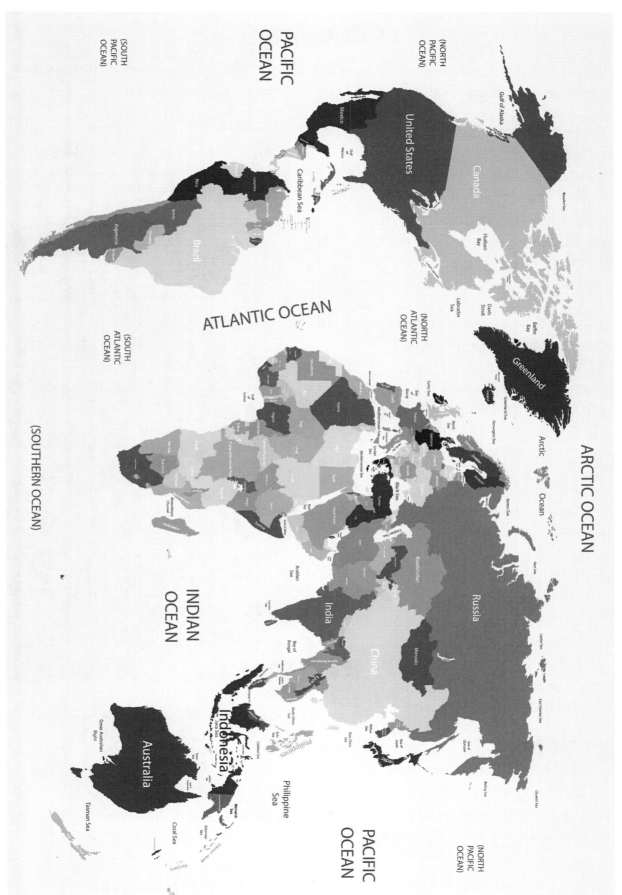

OTM2159   ISBN: 9781770783515
© On The Mark Press

# Glaciers

Glaciers are masses of ice and snow that have built up over time. They store large amounts of fresh water, in the form of ice and snow.

The lower parts of glaciers turn into ice because of the weight of the snow on top of them. The weight causes the lower part of the glacier to slowly melt and ooze along its bottom edges, making its way back to the ocean to become part of the 'conveyer belt'.

The cold air that emanates from the continental glaciers near the Earth's poles cools off the surrounding area. This cold, dry air is part of one of Earth's prevailing wind zones known as the polar easterlies.

The extremely cold air at the poles creates a high pressure, which doesn't bring much precipitation. The average amount of precipitation at the South Pole is 15 mm a year. The average amount of precipitation at the North Pole is 270 mm a year, with the wettest month being July with an average precipitation amount of 40 mm.

This is an alpine glacier at Glacier Bay Alaska. Alpine glaciers are found at high altitudes in the mountains.

This is a continental glacier on Iceland. Continental glaciers are found on the lands near the Earth's poles.

## Did You Know?

Antarctica is a continental glacier that lies over land. The Arctic is not a continent, but it is actually a frozen ice pack that lies over water, called the Arctic Ocean.

OTM2159   ISBN: 9781770783515
© On The Mark Press

# Glaciers and Sea Level

The source of ice in the ocean is an ice pack floating in the water and glaciers on land pushing in new ice. The Arctic Ocean's mass of an ice is an example of ice pack. Greenland and Antarctica are examples of glaciers. Global warming is causing more ice to melt.

**Question:** What type of melting ice is impacting sea level more: the Arctic ice pack or glaciers?

**You'll need:**

- a cutting board
- ice cubes
- access to tap water
- plasticine
- a clear rectangular container (about 9" x 13")
- a tablespoon
- a ruler
- salt

**What to do:**

1. Make a prediction to the question. Record it on worksheet 10.

2. Set up the cutting board in the container, using plasticine, as shown in the diagram.

3. Fill the container with water, so that the water is about 1.5 cm deep.

4. Add 1 tbsp. of salt to the water and stir around carefully with your hand. Allow the water to settle.

5. Fill the water with ice cubes, completely covering the board with them. (Some of the ice cubes will be out of the water). This is your glacier.

6. Using the ruler, measure the starting water depth. Record the time and the depth in the table on worksheet 10.

7. Continue to record the water depth every 10 minutes. Also, whenever an ice cube (this is an iceberg) slides into the water, record the time and measure the depth. Put an asterisk (*) beside to indicate that an iceberg has slid into the water.

8. When all of the ice has melted, graph your data on worksheet 11.

9. Make conclusions and a connection about what you have observed.

## Let's Predict

What type of melting ice is impacting sea level more: the Arctic ice pack or glaciers?

_____

_____

_____

## Let's Observe

Record the depth of water every 10 minutes, and when ice cubes fall into the water.

| TIME | WATER DEPTH |
|---|---|
|  |  |
|  |  |
|  |  |
|  |  |
|  |  |
|  |  |
|  |  |
|  |  |
|  |  |
|  |  |
|  |  |

OTM2159   ISBN: 9781770783515
© On The Mark Press

Name:

In the space below, graph the data that you collected. Be sure to give your graph a title and provide labels.

OTM2159   ISBN: 9781770783515
© On The Mark Press

## Let's Conclude

What happens to the water level over time?

_____

_____

_____

Circle the part of your graph where the biggest change in water level occurred. Compare this amount of change to what happened at the start. Use the data that you collected to explain the difference.

_____

_____

_____

_____

_____

_____

What type of melting ice is impacting sea level more: the Arctic ice pack or glaciers? Explain. _____

_____

_____

_____

_____

## Let's Connect It!

As global warming continues, which of the Earth's poles will impact the ocean levels the most? Explain your thinking.

_____

_____

_____

_____

_____

OTM2159   ISBN: 9781770783515
© On The Mark Press

# WASHING IN, WASHING OUT!

## LEARNING INTENTION:
Students will learn about the causes of tides, their range, and their effects on shorelines.

## SUCCESS CRITERIA:
- research tides and use a diagram to illustrate the cause of them
- describe two types of tides, illustrating the position of the Earth, sun, and moon
- describe and illustrate the meaning of the range of a tide
- list some possible effects of a tide on a shoreline
- identify the intertidal zone
- illustrate the intertidal zone and types of life forms that live in it

## MATERIALS NEEDED:
- a copy of "The Tides" worksheet 1, 2, 3, and 4 for each student
- a copy of "Affecting the Range" worksheet 5 and 6 for each student
- a large plastic bin (deep and rectangular), sand, rocks, gravel, a jug, a garden shovel, (for each pair of students)
- assorted small pieces of wood, plastic containers
- access to water
- access to the internet
- chart paper, markers
- clipboards, pencils

## PROCEDURE:
**\*This lesson can be done as one long lesson, or done in two or three shorter lessons.**

1. Give students worksheets 1, 2, 3, and 4. With access to the internet, they will research the causes of tides, types of tides, the range of a tide, and the location and life forms in the intertidal zone.

2. Divide students into pairs, and give them worksheets 5 and 6. Instruct them to design an experiment that will help them to investigate the factors that could possibly affect the range of a tide. Students will need access to assorted materials as they plan their investigation (see a list of ideas under the "Materials Needed" section above). *A follow-up option is to have pairs of students group with other pairs to observe the effects of their design. This would allow for discussion to occur using scientific terms and related topic vocabulary.*

## DIFFERENTIATION:
Slower learners may benefit by working in a small group with teacher support to complete the research about tides. A section of the research project could be assigned to each student in the small group, so that all sections are filled in to complete one final project. Each of their sections could be done on large chart paper, and then displayed in the classroom. An additional accommodation for these learners is to work with a strong peer to discuss the design of the tidal range effects investigation. Then, work together to plan and conduct the experiment.

**For enrichment**, faster learners could choose a tidal feature to research and describe, such as a tidal bore, a red tide, a rip tide, or a tsunami. The completed assignments could be shared with large group, and then displayed on a bulletin board.

# The Tides

Daily rises and falls of the ocean's waters are known as tides. When water rises to its highest level, it covers most of the shore. This is a high tide. When the water falls to its lowest level, it is at low tide. Conduct some research to find out more about tides.

## Let's Research!

Provide a written description and diagram to explain what causes a tide to occur.

_____

_____

_____

_____

_____

_____

_____

_____

OTM2159   ISBN: 9781770783515
© On The Mark Press

Name:

Illustrate the position of the moon, the sun, and the Earth during a **spring tide**.

Explain what happens during a spring tide.

_____

_____

_____

_____

Illustrate the position of the moon, the sun, and the Earth during a **neap tide**.

Explain what happens during a neap tide.

_____

_____

_____

_____

OTM2159   ISBN: 9781770783515
© On The Mark Press

## What is a tidal range?

_____

_____

_____

Use a diagram to illustrate the meaning of a tidal range.

List some possible effects of a tide on a shoreline:

- _____
- _____
- _____
- _____
- _____
- _____
- _____
- _____
- _____

OTM2159  ISBN: 9781770783515
© On The Mark Press

# Entering the intertidal zone...

## Where is it?

_____

_____

_____

## What does it look like? What lives there?

Name: _____

# Affecting the Range

You have learned a lot about tides. Your task now is to design an experiment to show what factor(s) could affect the range of a tide.

Formulate a question, make a list of materials, detail the procedure of your experiment, predict a result, conduct the experiment, make observations, and provide conclusions.

**Question:** _____

_____

_____

**You'll need:**

_____          _____

_____          _____

_____          _____

_____          _____

_____          _____

_____          _____

**What to do**

1. _____

2. _____

3. _____

4. _____

5. _____

6. _____

7. _____

8. _____

9. _____

10. _____

OTM2159   ISBN: 9781770783515
© On The Mark Press

Name:

## Let's Predict

_____

_____

_____

## Let's Observe

Illustrate and describe your observations.

Observations:

_____

_____

_____

_____

## Let's Conclude

_____

_____

_____

_____

_____

OTM2159   ISBN: 9781770783515
© On The Mark Press

# EROSION AND DEPOSITION

## LEARNING INTENTION:
Students will learn about displacement and deposition of sand, and the techniques used to prevent erosion of shorelines.

## SUCCESS CRITERIA:
- investigate and describe the effects of wave action on a coastline
- determine the meaning of vocabulary words associated with river development
- recognize the eroding power of a meandering river
- research, then compare and contrast the characteristics of young and mature rivers
- design, create, and test a shoreline erosion control system

## MATERIALS NEEDED:
- a copy of "At the Shoreline" worksheet 1 and 2 for each student
- a copy of "Rafting Down the River Vocab!" worksheet 3 for each student
- a copy of "Young vs. Mature Rivers" worksheet 4 for each student
- a copy of "Protecting the Shoreline" worksheet 5 for each student
- a copy of "Save Your Shoreline!" worksheet 6 and 7 for each student
- a copy of "Reporting From the Shoreline" worksheet 8 for each student
- a deep paint roller tray, a large pail of sand, a jug, a garden shovel, a pencil (for each group of students)
- access to water
- access to the internet
- Bristol board (a quarter of a piece for each pair of students)
- assorted small pieces of wood, sand, rocks, gravel, large rectangular plastic bins

- small plants, plant tray holders (from any garden centre)
- chart paper, markers, clipboards, pencils, rulers, compasses (mathematical)

## PROCEDURE:
*This lesson can be done as one long lesson, or divided into four or five shorter lessons.*

1. Give students worksheets 1 and 2, and the materials to conduct the experiment. Read through the question, materials needed, and what to do sections with students to ensure their understanding of the task. They will investigate how wave activity affects the shoreline of a beach, then record their observations and a conclusion on Worksheet 2. Upon completion, discuss with students the meaning of 'displacement' and deposition'.

2. Give students worksheet 3. With access to the internet or a dictionary, they will find the meaning of words associated with river development, and write a definition for each.

3. Give students worksheet 4. With access to the internet, they will research the characteristics of young and mature rivers. They will compare and contrast using the Venn diagram.

4. Give pairs of students a piece of thin cardboard or Bristol board, measuring about 25 cm x 25 cm. They will cut out a circular disk that measures 24 cm in diameter, and then mark the exact centre point of the circle. They will draw four inside circles on the cardboard (like a bull's eye), where the innermost circle has a radius from the centre point of 2.5 cm, the next circle has a radius of 5 cm from the centre point , the next circle has a radius of 7.5 cm, and the outermost circle has a radius of 10 cm. Using a compass, students will create the circles. They will mark a number 1 on the line of innermost circle, a number 2 on the

OTM2159  ISBN: 9781770783515
© On The Mark Press

second circle, a number 3 on the third circle, and a number 4 on the outermost circle. Using a sharp pencil, they will poke a hole through the centre of the disk. Placing a pencil through the centre, one partner will hold the pencil on either side of the disk, while the other partner spins it. Ask students to observe which numbers appear to be moving faster. (The outside numbers appear to move faster.)

**Questions for follow up discussion:**

- Does a meandering river flow faster on the inside or outside bank?

- Does a river deposit more sediment on the inside or outside bank as it meanders?

- Why does the outside of a meandering river erode faster than the inside bank? (*Meandering rivers flow slower on the inside banks and faster on the outside banks. Meandering rivers erode sediment from the outer bank, and deposit it on an inside bank further downstream, so there is more sediment on the inside bank.*)

- Would it be better to build a house on the inside or on the outside of a meandering river bank? Explain your thinking.

5. Give students worksheet 5. Read and discuss the concepts as a large group.

6. Explain to students that they will create a shoreline erosion control system. Divide students into pairs or small groups. Give them worksheets 6 and 7, and access to materials they can use to construct a system that will control shoreline erosion. (Systems created may be for a river or ocean setting, just as long as the system helps to control erosion of a shoreline. Be sure to have small natural and man-made materials on hand. Large, rectangular plastic bins could be used for ocean floors, and plant tray holders could be used to create river systems.)

7. Give students worksheet 8 to complete. *This 'inspection report' could be completed by students to review their own system, or it could be completed by a partner as an interview activity/review of another student's system.

## DIFFERENTIATION:

Slower learners may benefit by working together as a small group to complete worksheet 3. Each member could be assigned a vocabulary word to research the meaning of, and then bring it back to the small group. Responses could be written on a large piece of chart paper. This could be posted in the classroom as reference material for later activities. An additional accommodation is to follow this same format to complete worksheet 4 as a small group. Students could research some responses for the Venn diagram, and then compile their responses on a chart paper, which could be displayed in the classroom as well.

**For enrichment,** faster learners could choose a river in the world to research its history and characteristics. Product could be delivered in an oral/visual presentation about the river, or in a short essay format.

# At the Shoreline

Have you ever stood at the edge of a body of water, and watched as the waves washed up onto the shoreline? Let's conduct a simulation of this event!

**Question:** How does wave action affect a beach?

**You'll need:**

- a pencil
- a deep paint roller tray
- a garden shovel
- access to tap water
- a jug
- a large pail of sand

**What to do:**

1. Make a prediction to the question. Record it on worksheet 2.

2. Cover the bottom of the paint tray with sand. Create a beach at the shallow end of the tray.

3. Fill the jug of water and pour it into the deep end of the tray. (You may need to do this more than once, but be careful not to cover the beach.)

4. Observe the appearance of the beach before it is affected by wave activity. Record your observation on worksheet 2.

5. Place the pencil in the deep end of the tray. Make waves in the water by moving the pencil up and down, pushing it a few centimetres below the water each time.

6. Observe the appearance of the beach after it has been hit by wave action. Record your observation on worksheet 2.

7. Make a conclusion about what you have observed.

OTM2159  ISBN: 9781770783515
© On The Mark Press

## Let's Predict

How does wave action affect a beach?

_____

_____

_____

## Let's Observe

Describe the appearance of the beach before it is affected by wave activity.

_____

_____

_____

Describe the appearance of the beach after it was affected by wave activity.

_____

_____

_____

## Let's Conclude

_____

_____

_____

_____

### Displacing and depositing

Waves bring sand toward a beach, but as they retreat, the waves carry more sand away, displacing it from the shoreline then depositing it into the water.

# Rafting Down the River Vocab!

Access the internet or a dictionary to find out the meaning of vocabulary words associated with river development. Write a definition for each of the words below.

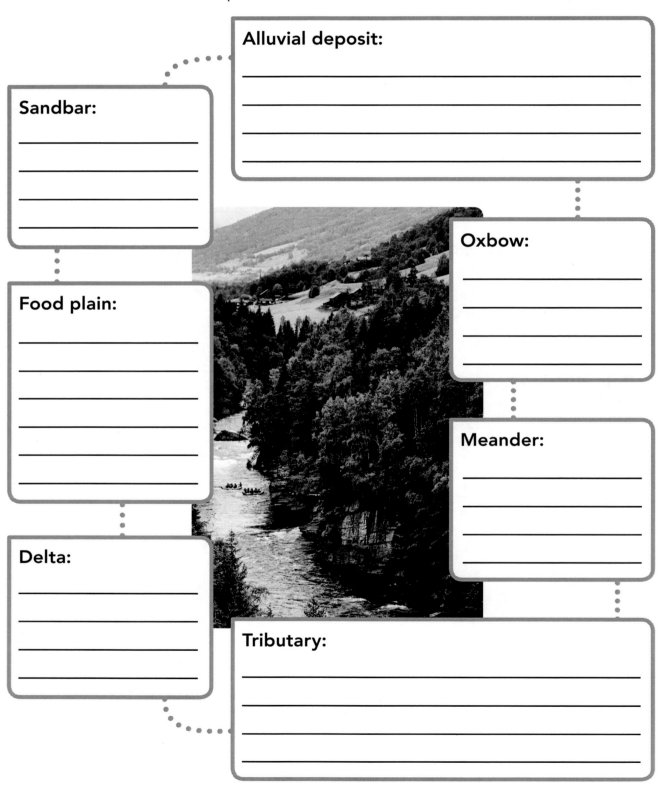

**Alluvial deposit:**

_____

_____

_____

_____

**Sandbar:**

_____

_____

_____

_____

**Oxbow:**

_____

_____

_____

_____

**Food plain:**

_____

_____

_____

_____

_____

**Meander:**

_____

_____

_____

_____

**Delta:**

_____

_____

_____

_____

**Tributary:**

_____

_____

_____

_____

_____

58

OTM2159   ISBN: 9781770783515
© On The Mark Press

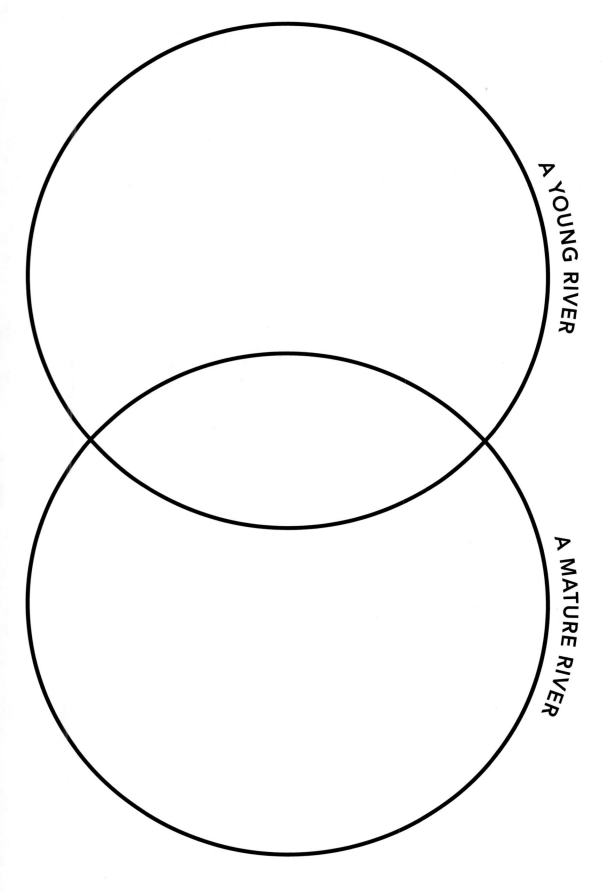

# Young vs. Mature Rivers

Your task now is to do some research on the characteristics of young and mature rivers. Use the Venn diagram below to compare and contrast. Remember the middle of the Venn diagram is used to list similarities!

A YOUNG RIVER

A MATURE RIVER

OTM2159　ISBN: 9781770783515
© On The Mark Press

# Protecting the Shoreline

Fighting the forces of Mother Nature can be challenging! Tides will continue to wash in and out, and rivers will continue to steadily flow, causing erosion of our land's edges. There are, however, many examples of human intervention to prevent riverbank or coastal erosion. Let's learn about some of them!

The presence of vegetation around the shoreline of coastal areas and riverbanks helps to prevent erosion. Some people plant vegetation because it acts as a natural barrier against forces such as wind and moving water from eroding the land. Marram Grass is an example of a tall perennial grass that grows healthily on seaward slopes to prevent sand from being washed away, and helps to build sand dunes.

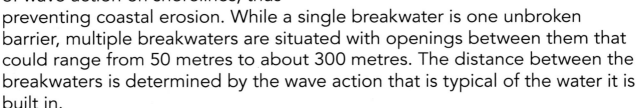

Breakwaters are structures that are put in place to reduce the intensity of wave action on shorelines, thus preventing coastal erosion. While a single breakwater is one unbroken barrier, multiple breakwaters are situated with openings between them that could range from 50 metres to about 300 metres. The distance between the breakwaters is determined by the wave action that is typical of the water it is built in.

Breakwaters can be constructed parallel or perpendicular, depending on the shoreline requirements.

Breakwaters are usually stationed offshore, ranging from 100 metres to about 500 metres from the shoreline.

OTM2159   ISBN: 9781770783515
© On The Mark Press

# Save *Your* Shoreline!

Design and construct your own system that will protect a shoreline of a body of water. Be sure to test it to determine its functionality!

**The materials I will use are:**

_____

_____

_____

_____

_____

_____

_____

_____

_____

**A design of my erosion control system:**

**My plan for making the erosion control system is:**

1. _____

2. _____

3. _____

4. _____

5. _____

6. _____

7. _____

8. _____

9. _____

10. _____

**Now it's time to create a 'before' and 'after'! Go to worksheet 7.**

## *The Before...*

Illustrate and describe your body of water's movement in terms of flow rate or wave action, before adding in an erosion control system. What effect is it having on the land?

_____

_____

_____

_____

## *The After...*

Illustrate and describe your body of water's movement in terms of flow rate or wave action, after adding in an erosion control system. What effect is it having on the land?

_____

_____

_____

_____

OTM2159  ISBN: 9781770783515
© On The Mark Press

Name:

# Reporting From the Shoreline

Complete an inspection report of your shoreline. Review the erosion control techniques used to protect your shoreline by answering the following questions:

1) What function was your erosion control system intended to perform?

_____

_____

2) What strategies did you use to ensure you had a design that could withstand or lessen the effects of flow rate or wave action?

_____

_____

_____

_____

3) How did you test your erosion control system?

_____

_____

_____

_____

4) Was your system successful in erosion control? How do you know?

_____

_____

_____

_____

5) What changes might you suggest to improve its design and functionality?

_____

_____

# YOUR WATER SUPPLY

## LEARNING INTENTION:

Students will learn about where water comes from, the water treatment process, their usage of water, and ways to reduce water consumption.

## SUCCESS CRITERIA:

- determine where water is extracted from
- identify factors that can affect the water table
- investigate and describe the local water treatment process
- recognize the amount of water required to do some daily living activities
- make observations and record results about your family's use of water
- analyze and compare personal water consumptions to that in varying countries
- make connections to the environment as you look at ways to reduce water usage

## MATERIALS NEEDED:

- a copy of "The Water Table" Worksheet 1 for each student
- a copy of "Water Table Talk!" Worksheet 2 for each student
- a copy of "From the Ground to Our Taps!" Worksheet 3 for each student
- a copy of "Investigating the Local Water Supply" Worksheet 4, 5, 6, and 7 for each student
- a copy of "How Much Water Do You Use?" Worksheet 8, 9, 10, and 11 for each student
- a copy of "Reduce Your Use!" Worksheet 12 for each student
- access to the internet
- clipboards, pencils, chart paper, markers

## PROCEDURE:

*This lesson can be done as one long lesson, or divided into four or five shorter lessons. Items 1, 2, 3, 5, 7, 8, 9, and 10 can be done at school. Item 4 requires students to visit a water treatment facility to complete this component. Item 6 contains a homework component.

1. Give students Worksheet 1. Read through with the students about the water table. Along with the content, discussion of certain vocabulary words would be of benefit to ensure students' understanding of the concepts.

   **Some interesting vocabulary words to focus on are:**

   - evaporation
   - saturated
   - aquifer
   - impermeable
   - condensation
   - unsaturated
   - water table
   - penetrate
   - precipitation
   - zone of saturation
   - extract
   - artesian well

2. Divide students into partners and give them Worksheet 2. Students will think about and discuss natural and human factors that could affect the level of the water table. An option is to following this up with a large group discussion, recording student responses in a T-chart, with headings of 'natural factors' and 'human factors'. (Some possible responses are droughts, rainy season, snowmelt, presence of swamp vegetation, floods, tidal tables, lawn watering, overuse of wells, inefficient showers and toilets, impervious surfaces such as concrete prevents water from entering the saturation zone and becomes run off instead, extraction of water by bottle water companies.)

3. Give students Worksheet 3. Read through the water processing steps with students and discuss to ensure their understanding of the concepts.

4. **Arrange for students to take a tour of your local water treatment facility.** Give students

OTM2159  ISBN: 9781770783515
© On The Mark Press

a pencil, clipboard, and Worksheets 4, 5, 6, and 7, to take on the tour. Encourage students to ask questions to inquire about the extraction, cleaning process of their local water supply, and how waste water is managed. Upon return, have a discussion about the things that they learned, what they think others in the community should know, and how they think the message should be passed on to the community about their local water supply and usage.

5. Give students Worksheet 8. Read and discuss the information as a large group to ensure students' understanding of the information about how we use water in our daily lives, and how much of it we use. Students can complete the question at the bottom of the Worksheet, individually, or as a 'turn and talk' exercise with a peer.

6. Explain to students that they will conduct an investigation about their family's water usage around the house. Give them Worksheet 9. Explain that they will use this chart at home to record **how many times** water is used by family members in their house. A class discussion about how else water is used in the home may be of benefit, as this may provide guidance to students' choices of 'other' in the chart.

7. After students have had **2 days** to complete Worksheet 9, ask them to return it to school so that they can assess the data that they collected. Give students Worksheet 10. Students will count up the number of squares they colored in for each activity on Worksheet 9 and record this on Worksheet 10. Referring back to Worksheet 8, students will record the litres of water required to do each of the activities listed on Worksheet 10. Using a calculator, they will multiply the number of times the activity was done by the number of litres of water it takes to do it each time. Then they will record this information in the chart on Worksheet 10.

8. Give students Worksheet 11. They will use the data that they collected about their families' water usage to analyze, assess, and compare their personal water consumption to that of some other countries. An option is to come back together as a large group to discuss their findings and ideas.

9. Have a large group discussion about the following ideas:
   • What does it mean to use water excessively?
   • What does it mean to use water efficiently?
   • How might our water usage change if we had to carry our water into our homes instead of just turning on the tap?

10. Divide students into pairs and give them Worksheet 12. Students will engage in a 'Think-Pair-Share' activity in which, they will discuss and respond about responsible water use. An option is to come back together as a large group to discuss and record ideas on chart paper.

## DIFFERENTIATION:
Slower learners may benefit by working in a small group with teacher support to complete Worksheets 10 and 11. This will ensure that the data collected is calculated correctly so that students can then work to analyze and assess their data. These learners may also benefit by only recording their own thinking on Worksheets 2 and 12, and only listening to their partner's responses that are shared.

**For enrichment**, faster learners could create a bar graph that depicts their data collected for their family's water usage. An alternative or additional enrichment activity for these learners would be to have them write a newspaper article (letter to the editor) that informs and persuades others to reduce their water usage.

# The Water Table

You know that in the Earth's water cycle, evaporation from the ground rises up through the air, where it condenses in the clouds, then falls as precipitation back to the ground. But, does it all get evaporated? Let's learn more about this!

As precipitation falls, it penetrates into the Earth's ground soil. While some of this precipitation gets evaporated, some of it seeps through the layers of ground soil, leaving it saturated and allowing a water table to form.

A **water table** is described as the area between unsaturated ground and water-saturated ground. The water table sits on a **zone of saturation**, which is an area of rock and soil that is full of water.

There are pockets of water that exist below the water table, which are called aquifers. **Aquifers** are used to extract water for human usage. People living in a rural area, have water wells on their land. Water wells can be dug below the water table and the water pumped out for human usage.

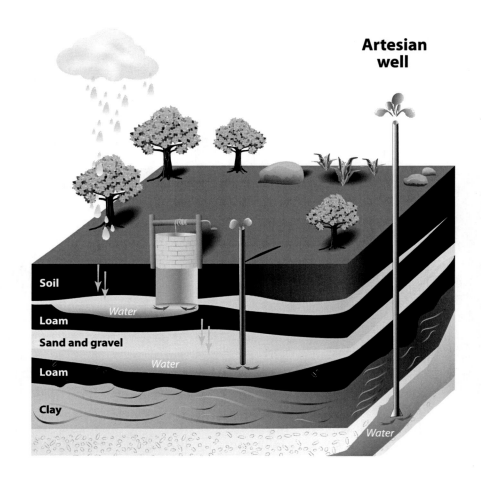

An **artesian well** has water steadily flowing up from it, sometimes without the use of a pump. As water reaches an impermeable layer of rock, above and below it, the pressure of the water's weight forces it to the surface from a well drilled down into the aquifer. Artesian wells are a major source of water in places adjacent to mountain ranges that receive precipitation.

OTM2159  ISBN: 9781770783515
© On The Mark Press

Name:

# Water Table Talk!

With a partner, do some thinking and talking about the questions below. Record your ideas in the chart.

### "What are some natural factors that could cause a change in the water table level?"

| My Thinking | My Partner's Thinking |
| --- | --- |
| | |
| | |
| | |
| | |
| | |
| | |
| | |

### "What are some human factors that could cause a change in the water table level?"

| My Thinking | My Partner's Thinking |
| --- | --- |
| | |
| | |
| | |
| | |
| | |
| | |
| | |

OTM2159   ISBN: 9781770783515
© On The Mark Press

# From the Ground to Our Taps!

You know that you need water. You know that you use it every day. Have you ever wondered how you get it to your tap? Let's take a look at the water treatment process!

Water is brought into a water treatment plant from a natural source like a lake. Once the water is in the plant, it goes through many steps to be cleaned and ready to come out of your tap in your home.

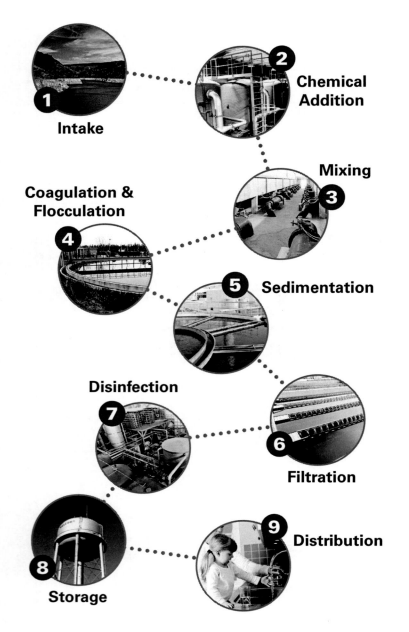

**Steps:**

1. Water is taken from a natural source, like a lake.

2. Chemicals are added to the water to kill germs.

3. The chemicals and the water get mixed together.

4. Particles stick together to form flocculants or "floc."

5. The water and floc go into a basin. The floc sinks to the bottom and is taken out of the water.

6. Water flows through filters.

7. More chemicals are added to kill any other germs.

8. Water is stored in a tank.

9. Water is taken to houses and businesses.

OTM2159  ISBN: 9781770783515
© On The Mark Press

# Investigating the Local Water Supply

You will now use your investigative/research skills to find out more about the water supply in your community. Take a tour of your local water treatment facility. Use the questions below as a guideline for your research.

## Going On Tour!

What body of water does the water supply for your community come from?

_____

_____

_____

How is the water extracted?

_____

_____

_____

_____

List the steps involved to process the water so that it is ready for human usage.

- _____
- _____
- _____
- _____
- _____
- _____
- _____
- _____
- _____

Name the chemicals that are added to the water, *and* explain why they are added.

_____

_____

_____

_____

_____

_____

Draw a detailed diagram on one of the processing steps that you listed on Worksheet 4. **(Your diagram must include labels that provide information about the illustration.)**

OTM2159　ISBN: 9781770783515
© On The Mark Press

How is the water distributed to local consumers?

_____

_____

_____

_____

_____

How is consumption measured?

_____

_____

_____

_____

_____

How is waste water collected?

_____

_____

_____

_____

_____

How is waste water managed?

_____

_____

_____

_____

_____

OTM2159   ISBN: 9781770783515
© On The Mark Press

What variables could possibly affect the local water supply? How?

_____

_____

_____

_____

_____

_____

_____

How would these be addressed by your city, town, or region?

_____

_____

_____

_____

_____

_____

_____

_____

## Post Tour...

In *your* opinion, what should people know about their local water supply and how it is managed?

_____

_____

_____

_____

_____

_____

OTM2159   ISBN: 9781770783515
© On The Mark Press

Name:

# How Much Water Do You Use?

In the chart below are some examples of how we use water in our homes and how much of it we use.

**Did You Know?**

| Daily activities that use water | Average amount of litres of water it takes to do it |
| --- | --- |
| Brushing your teeth | 1 litre |
| Flushing the toilet | 6 litres |
| Washing your hands and face | 7 litres |
| Taking a bath | 80 litres |
| Taking a **5-minute** shower | 35 litres |
| Running the dishwasher | 20 litres |
| Washing a sink of dishes by hand | 30 litres |
| Using the clothes washing machine | 70 litres |
| Drinking water per day | 1 litre |

What are some other ways that we use water in and around our homes?

_____

_____

_____

_____

_____

Name:

## Let's Collect Data

Record how often water is used in your home by your family for **2 days**. Color in one square in the correct section each time it is used.

| Brushing teeth | | | | | | | | | | | | | | | | | | | | | | |
|---|---|---|---|---|---|---|---|---|---|---|---|---|---|---|---|---|---|---|---|---|---|---|
| Flushing toilet | | | | | | | | | | | | | | | | | | | | | | |
| Washing hands or face | | | | | | | | | | | | | | | | | | | | | | |
| Taking a bath | | | | | | | | | | | | | | | | | | | | | | |
| Taking a 5-minute shower | | | | | | | | | | | | | | | | | | | | | | |
| Running the dishwasher | | | | | | | | | | | | | | | | | | | | | | |
| Hand washing a sink of dishes | | | | | | | | | | | | | | | | | | | | | | |
| Using the washing machine (clothes) | | | | | | | | | | | | | | | | | | | | | | |
| Drinking water per person | | | | | | | | | | | | | | | | | | | | | | |
| Other? _____ | | | | | | | | | | | | | | | | | | | | | | |
| Other? _____ | | | | | | | | | | | | | | | | | | | | | | |

OTM2159   ISBN: 9781770783515
© On The Mark Press

Name:

## Let's Calculate

Calculate how much water, was used by your family for 2 days. You will need to look at the information on Worksheet 8 and the data that you collected on Worksheet 9.

| Activities using water | Number of squares colored in the chart | Number of litres of water it takes to do the activity *one time* | Amount of litres of water used to do the activity over *2 days* |
|---|---|---|---|
| Brushing teeth | | | |
| Flushing toilet | | | |
| Washing hands or face | | | |
| Taking a bath | | | |
| Taking a 5-minute shower | | | |
| Running the dishwasher | | | |
| Hand washing a sink of dishes | | | |
| Using the washing machine (clothes) | | | |
| Drinking water per person | | | |
| Other? _____ | | | |
| Other? _____ | | | |

## Analyze the Data!

Look back at the chart on Worksheet 10 and analyze the data that you have collected.

1) In total, how many litres of water did your family use over 2 days?

_____

_____

_____

2) Calculate the average amount of water used by each member of your family.

Personal water consumption over **2 days** = _____

Personal water consumption **per day** = _____

3) Access the internet to find out the average personal water consumption per day in other countries.

Personal water consumption per day
in one of my neighboring countries = _____

Personal water consumption per day
in a country on another continent = _____

4) How does your family's average personal water consumption per day compare to the average personal water consumption in some other countries?

_____

_____

_____

_____

_____

_____

_____

_____

_____

OTM2159　ISBN: 9781770783515
© On The Mark Press

Name:

# Reduce Your Use!

With a partner, do some thinking and talking about the questions below.

## "Why shouldn't we waste water?"

| My Thinking | My Partner's Thinking |
|---|---|
| | |

## "How is your family *already* showing a responsible use of water?"

| My Thinking | My Partner's Thinking |
|---|---|
| | |

## "What are some other ways your family could reduce their water usage in the future?""

| My Thinking | My Partner's Thinking |
|---|---|
| | |

OTM2159   ISBN: 9781770783515
© On The Mark Press

# NATURE'S WATER

## LEARNING INTENTION:

Students will learn about obtaining drinking water in areas where supply is limited; and assess how media sources address issues related to the sustainability of the Earth's water systems.

## SUCCESS CRITERIA:

- create fresh water from salt water
- construct a filtration system
- determine the causes of pollution
- describe the environmental and societal impacts of pollution in our water
- research and assess how a media source addresses the issue of pollution and the sustainability of our water systems

## MATERIALS NEEDED:

- a copy of "Nature's Drink of Water" Worksheet 1 and 2 for each student
- a copy of "Pollution and Our Water" Worksheet 3 and 4 for each student
- a copy of "According to the Media" Worksheet 5 and 6 for each student
- a large bowl, a spoon, a cup/ glass, a rock (for each student)
- a bucket of water, salt, plastic wrap, a few rolls of masking tape, a few large bowls
- a sunny day
- a large clear plastic bottle, a cup/ glass (one per student)
- a few hammers and large nails, a few pairs of scissors (or knives), a few shovels, charcoal (burnt wood pieces)
- a kettle/ pot over a campfire (optional)
- access to a river or stream with the availability of assorted sizes of rocks, sand, grass

- access to media sources such as newspapers, magazines, television, radio, or internet that address issues of water pollution and the effects of human activities on the long-term sustainability of a water system
- chart paper, markers, clipboards, pencils

## PROCEDURE:

*This lesson can be done as one long lesson, or done in four or five shorter lessons. Item 2 requires students to visit a riverbank or stream to complete this component.

1. On a sunny day, give students Worksheet 1, and access to the materials to conduct the experiment on desalinating salt water. (The longer the water is left in the sun, the more water will accumulate in the glass.) As students explain how the drink of fresh water was created, ask them to reflect back on the information they learned about the water cycle, and to relate their answer to their knowledge of scientific terms of the water cycle.

2. **Take students to a nearby river bank or stream.** Give them Worksheet 2, a clipboard and pencil, and access to the materials to build and test a water filtration device. (To save time, or in the interest of safety, pre-cut the plastic bottles and make holes in the bottle caps.) As students explain how the drink of water was cleaned, ask them to reflect back on the information they learned about the water treatment process, and to relate their answer to their knowledge of scientific terms of this process.

3. Divide students into pairs and give them Worksheets 3 and 4. Students will engage in a 'Think-Pair-Share' activity in which, they will discuss and respond about pollution and our water. An option is to gather as a large group to discuss and record ideas on chart paper.

OTM2159  ISBN: 9781770783515
© On The Mark Press

4. Explain to students that they will need to access and assess a publication that addresses issues of water pollution and the effects of human activities on the long-term sustainability of a water system. The media source they could access for this information could be the newspaper, a magazine article, a television news piece, a radio broadcast, the internet, or a documentary. Give students a clipboard and pencil, and Worksheets 5 and 6. *A follow-up option of this activity is to put students into small groups of 3 or 4, so that they can share their publications and the messages behind them. If some students have information on the same topic, but accessed it from a different source, encourage them to compare the information and discuss their findings and opinions.*

## DIFFERENTIATION:

Slower learners may benefit by working in a small group with teacher support to complete Worksheets 5 and 6. A media piece could be pre-selected to ensure that it is suited to the interest, learning style, and level of the group.

**For enrichment**, faster learners could research and assess a scientific technology that is currently affecting the Earth's water systems, in terms of the impact the advances in the technology is having on our water systems (e.g., bioremediation, desalination).

# Nature's Drink of Water

What would **you** do if you did not have access to drinking water from a tap or from bottled water? How could you get it if you were in an area where supply is limited, or could not be found?

## Scenario #1 – an ocean setting

Try this!

Using 2 bowls, water, salt, a spoon, plastic wrap, masking tape, a cup, a rock, and a sunny day, get yourself a drink of water!

- add salt to a bowl of water, stir until it is dissolved (this is ocean water)
- pour the salt water into another bowl so that it is about 5 cm deep
- place a glass into the centre of the bowl (the glass must be higher than the water level, but lower than the height of the bowl)
- cover the top of the bowl with plastic wrap, and seal the edge with tape
- place the rock on top of the plastic wrap, directly above the glass that is inside the bowl
- place the bowl and its contents in the sun for a whole day (or longer)
- remove the plastic wrap, and there should be a glass of water waiting for you!

Use scientific terminology to explain how your drink of fresh water was created.

_____

_____

_____

_____

_____

_____

_____

_____

_____

_____

_____

**Diagram:**

OTM2159   ISBN: 9781770783515
© On The Mark Press

# Scenario #2 – at the river bank

Using a large plastic bottle with a cap, a knife or scissors, a hammer and a nail, a glass, a shovel, rocks (small, medium, and large), sand, charcoal (burnt wood pieces), grass, and a kettle (or pot over a campfire), filter yourself a drink of water!

- using the hammer and nail, poke some holes into the cap of the bottle
- carefully cut off the bottom of the plastic bottle
- turn the bottle upside down and place a layer of large rocks into it
- place a layer of medium-sized rocks into the bottle
- place a layer of small rocks into the bottle
- put a layer of sand on top of the rocks, and pack it down
- put some charcoal on top of the sand layer, then add more sand
- put some grass on top of the sand
- use the shovel to dig an "Egyptian well" (a hole beside the water's edge); the water table will show itself, so use the glass to grab this water as it rises
- pour this river water through the filtration system that you created, quickly placing the glass underneath the perforated cap to catch the filtered water
- once the water has filtered through, filter it again
- there should be a glass of water waiting for you; boil it if you wish!

Use scientific terminology to explain how your drink of water was cleaned.

_____
_____
_____
_____
_____
_____
_____
_____
_____
_____
_____
_____

**Diagram:**

# Pollution and Our Water

With a partner, do some thinking and talking about the questions below. Record your ideas in the chart.

## "What substances are causing pollution of our waters?"

| My Thinking | My Partner's Thinking |
| --- | --- |
| _____ | _____ |
| _____ | _____ |
| _____ | _____ |
| _____ | _____ |
| _____ | _____ |
| _____ | _____ |
| _____ | _____ |

## "What are the related environmental impacts of this pollution in our waters?"

| My Thinking | My Partner's Thinking |
| --- | --- |
| _____ | _____ |
| _____ | _____ |
| _____ | _____ |
| _____ | _____ |
| _____ | _____ |
| _____ | _____ |
| _____ | _____ |

OTM2159  ISBN: 9781770783515
© On The Mark Press

With your partner, continue to think and talk about the questions below. Record your ideas in the chart.

### "What are the related societal impacts of this pollution in our waters?"

| My Thinking | My Partner's Thinking |
|---|---|
| _____ | _____ |
| _____ | _____ |
| _____ | _____ |
| _____ | _____ |
| _____ | _____ |
| _____ | _____ |
| _____ | _____ |
| _____ | _____ |

### "What are some ways to reduce or eliminate this pollution in our waters?"

| My Thinking | My Partner's Thinking |
|---|---|
| _____ | _____ |
| _____ | _____ |
| _____ | _____ |
| _____ | _____ |
| _____ | _____ |
| _____ | _____ |
| _____ | _____ |
| _____ | _____ |

OTM2159   ISBN: 9781770783515
© On The Mark Press

Name: _____

# According to the Media

Assess how a media source addresses the issue of water pollution and the effects of human activities on the long-term sustainability of a water system. Choose from the following media sources:

- a newspaper or magazine article
- an information piece on the internet
- a radio discussion broadcast
- a television news piece
- a documentary

## Assessing the Source!

**Title of publication:** _____

**Media source:** _____

What is the main idea of this publication?

_____

_____

_____

_____

State the supporting details in the piece.

- _____
- _____
- _____
- _____
- _____
- _____
- _____
- _____
- _____

OTM2159   ISBN: 9781770783515
© On The Mark Press

Are there any implied messages in the piece? Explain.

_____

_____

_____

_____

In your opinion, who is the intended audience for this piece?

_____

_____

_____

What different groups or opinions are represented in this piece?

_____

_____

_____

_____

In your opinion, how might different groups of people react to the messages in the piece?

_____

_____

_____

_____

_____

How has this piece influenced or inspired *you*? Explain.

_____

_____

_____

_____

OTM2159   ISBN: 9781770783515

# WATER TESTING

## LEARNING INTENTION:

Students will learn how to test a variety of water samples for presence of different chemicals; and assess the importance of this knowledge in terms of having healthy drinking water.

## SUCCESS CRITERIA:

- test different water samples for levels of pH, alkalinity, free chlorine, and nitrate
- collect and display data using charts
- make and record observations of the chemical levels in each water sample
- make and record conclusions about safe drinking water and harmful drinking water
- make a connection to the land and waterways in our environment

## MATERIALS NEEDED:

- a copy of "Putting It to the Test!" Worksheet 1, 2, 3, and 4 for each student
- 3 water testing kits (Premier water testing kits can be ordered on-line, or other brands can be purchased at hardware or pool care stores)
- 5 measuring cups, a timer
- a bottle of bottled water, a jar of tap water, a jar of filtered water
- a jar of rain water
- a jar of river, pond, stream, or lake water
- clipboards, pencils

## PROCEDURE:

1. Explain to students that they will test different water samples for a variety of chemicals. Give them Worksheets 1, 2, 3, and 4. Due to the price point of the water testing kits, conduct this investigation as a large group. Different students could be chosen to dip the testing strips and to read the levels, as the remaining students record the information in the data collection charts on Worksheet 2 and 3. Allow students some time to record their observations after each chemical test is completed. On Worksheet 4, students will make conclusions and a connection about their findings.

## DIFFERENTIATION:

Slower learners may benefit by working in a small group with teacher support to discuss the questions in the conclusion and connection sections on Worksheet 4. This would provide guidance for them as they review their observations. An option is to eliminate the written component for them in lieu of this discussion.

**For enrichment**, faster learners could research, discuss, or debate a specific human practice or technology that could threaten the quality of surface or ground water systems in their area. For example, development of a mine, location of a new landfill site, building of a pulp and paper mill, or development of residences on a lake or river that is used as a water supply for the local population.

- what is the human practice or technology in question?
- what water system is potentially threatened in quality?
- what are the societal groups with different needs in relation to the practice or technology in question?
- how are the decisions and actions of these different groups being addressed?

OTM2159  ISBN: 9781770783515
© On The Mark Press

Name:

# Putting It to the Test!

Have you wondered what is in the water? If you put it to the test, you will discover that there are chemicals in water. Some amounts of certain chemicals are safe, and some are not. Let's put some water samples to the test!

## You'll need:

- a water testing kit
- a jar of filtered water
- a jar of tap water
- a jar of river, pond, stream, or lake water
- a jar of rain water
- a bottle of bottled water
- a timer
- 5 measuring cups

## What to do:

1. Pour 250 mL of the tap water into a measuring cup. Pour 250 mL of the filtered water into a measuring cup.

2. Pour 250 mL of the bottled water into a measuring cup. Pour 250 mL of the rain water into a measuring cup.

3. Shake up (or stir) the jar of river, pond, stream, or lake water. Pour 25 mL of it into a measuring cup.

4. Dip a pH strip into each water sample, for about 20 seconds.

5. Pull them out and take the reading of the pH level according to the chart that is in the water testing kit. Record the levels of each water sample in the chart on Worksheet 2, and make observations from the data that you collect.

6. Repeat steps 4 and 5 using the alkalinity testing strips.

7. Repeat steps 4 and 5 using the chlorine testing strips and the nitrate testing strips. Record levels for these categories for each water sample in the charts on Worksheet 3, and make observations from the data.

8. On Worksheet 4, make conclusions and connections about what you have observed.

## Let's Collect Data!

### pH Levels

A pH scale measures how acidic or basic a substance, like water, is. A pH level less than 7 is acidic, and a pH level greater than 7 is basic. A pH level of 7 is neutral, and this is what you want your drinking water to be.

| Water Samples | pH Levels |
|---|---|
| tap water | |
| filtered water | |
| bottled water | |
| rain water | |
| river, stream, pond, or lake water | |

Observations: _____

_____

_____

### Alkalinity Levels

Alkalinity is the ability of water to neutralize acidity. An alkalinity test measures the level of bicarbonates, carbonates, and hydroxides in water. Levels are measured as ppm (parts per million). Levels between 30 and 60 ppm are considered optimal.

| Water Samples | pH Levels |
|---|---|
| tap water | |
| filtered water | |
| bottled water | |
| rain water | |
| river, stream, pond, or lake water | |

Observations: _____

_____

_____

OTM2159   ISBN: 9781770783515
© On The Mark Press

## Chlorine Levels

Chlorine is a disinfectant that, when added to water, reduces or eliminates bacteria. If enough chlorine is added to water, some will stay in the water after all organisms have been destroyed. This free chlorine stays in the water until it is used to destroy new contamination. Most drinking water has free chlorine levels of 0.04 - 2.0 mg/L.

| Water Samples | pH Levels |
|---|---|
| tap water | |
| filtered water | |
| bottled water | |
| rain water | |
| river, stream, pond, or lake water | |

Observations: _____

_____

_____

## Nitrate Levels

Nitrate occurs naturally in groundwater. Activities, such as livestock operations or sewage disposal sites near a water supply, or areas with chemical fertilizer can contaminate the water. An acceptable level of nitrate in drinking water is 10mg/L.

| Water Samples | pH Levels |
|---|---|
| tap water | |
| filtered water | |
| bottled water | |
| rain water | |
| river, stream, pond, or lake water | |

Observations: _____

_____

_____

## Let's Conclude

According to your observations, which water sample(s) is the safest to drink? Explain your thinking.

_____

_____

_____

_____

_____

According to your observations, which water sample(s) could be harmful for humans to ingest? Explain your thinking.

_____

_____

_____

_____

_____

## Let's Connect It!

In the construction of new landfill sites, there is preparation to do before garbage is dumped in it. A hole is dug and a thick plastic lining is put down that collects the leachate, which is a liquid produced as waste decays. Leachate is then disposed of at a waste water treatment facility.

Explain the importance of this procedure in relation to our waterways and to the water that we drink.

_____

_____

_____

_____

_____

_____

_____

_____

OTM2159   ISBN: 9781770783515
© On The Mark Press

# THE AQUATIC ENVIRONMENT

## LEARNING INTENTION:
Students will learn about the productivity and distribution of life forms in aquatic environments, how changes are monitored, and the effects of human activity on their environments.

## SUCCESS CRITERIA:
- identify factors affecting the productivity and distribution of an aquatic species
- research and describe the habitat of an aquatic life form, its nutrient needs, and its predators
- determine the probability of the aquatic life form's abundance
- identify possible changes to the aquatic life form's environment and its adaptations
- explain how changes in the aquatic life form's environment are monitored
- recognize effects of human activity to the aquatic life form's productivity and distribution

## MATERIALS NEEDED:
- a copy of "What's Affecting Life in the Water?" Worksheet 1 for each student
- a copy of "In the Life of an Aquatic Species" Worksheet 2, 3, 4 and 5 for each student
- access to the internet, Bristol board (a piece for each student)
- chart paper, markers, clipboards, pencils, pencil crayons, glue, scissors

## PROCEDURE:
*This lesson can be done as one long lesson, or be divided into shorter research periods.*
1. Divide students into pairs. Give them Worksheet 1. They will engage in a 'turn-and-talk' activity where they will exchange and record ideas about the factors that could affect the productivity and distribution of species living in aquatic environments. Encourage them to relate their ideas to the knowledge they have gained about water (e.g., currents, pollution, temperature, water depth, sunlight, turbidity, salinity, resource extraction, fishing, dams). An option is to allow students to access the internet for ideas. Upon completion, come together as a large group to share ideas. Record ideas on chart paper for future reference.

2. As a large group, brainstorm and record examples of life forms found in fresh water, and those found in salt water. These examples can be any aquatic species, even plant life.

3. Give students Worksheets 2, 3, 4 and 5. Students will choose an aquatic species. They will access the internet to research its habitat, nutrient needs, predators, adaptations, how their environment is monitored, and the effects of human activity to their environment.

4. Upon completion of the research assignment, divide students into small groups. They will present the information on their chosen aquatic life form to the small group. *An alternative option is to have students display their information on Bristol board, then display it around the room so that students can participate in a carousel activity to read and learn about their classmates' chosen aquatic life forms.*

## DIFFERENTIATION:
Slower learners may benefit by working in a small group with teacher support to complete the research about one aquatic life form. A section of the research project could be assigned to each student in the small group, so that all sections are filled in to complete one final project. Each of their sections could be done on large chart paper or on one piece of Bristol board, and then displayed in the classroom.

**For enrichment**, faster learners could research historical Aboriginal information or interview an Aboriginal elder, to learn more about how aquatic resources are regarded by First Nations people, in terms of seasons, locations, methods of capture, the usages, and sustainability.

OTM2159   ISBN: 9781770783515
© On The Mark Press

# What's Affecting Life in the Water?

You have learned a lot about water. You know how it moves, the effects of its movement on the Earth's surface, its chemical makeup, and the pollutants that inhabit it. But how does water affect the organisms that are living in it?

**With a partner**, do some thinking and sharing of ideas about factors that could affect the productivity and distribution of species living in aquatic environments. Record your ideas below.

**Factors affecting species in aquatic environments:**

OTM2159   ISBN: 9781770783515
© On The Mark Press

Name:

# In the Life of an Aquatic Species

Choose an aquatic life form that you would like to know more about. Conduct some research to find out about:

- where the aquatic life form lives
- what it eats/how it gets its nutrients
- what are its predators
- how it adapts to changes in its environment
- what it contributes to its environment
- the effects of human activity to its productivity and distribution

**Let's Research!**

**Aquatic Life Form:** _____

Provide a written description and detailed diagram of its habitat.

_____
_____
_____
_____
_____

In the T-chart, make a list of the nutrients that your aquatic life form feeds on.
Then make a list of the predators that feed on your aquatic life form.

| Nutrients for the aquatic life form | Predators of the aquatic life form |
|---|---|
| | |

## Thinking critically...

How abundant is the species of your aquatic life form? (Think about the
availability of nutrients it feeds on vs. the amount and presence of its
predators.)

_____

_____

_____

_____

_____

OTM2159   ISBN: 9781770783515
© On The Mark Press

Name:

List some possible changes in its environment your aquatic life form might have to face.

- _____
- _____
- _____
- _____
- _____
- _____

Choose two of the changes that you have listed above. How would your aquatic life form adapt or respond to these changes in its environment?

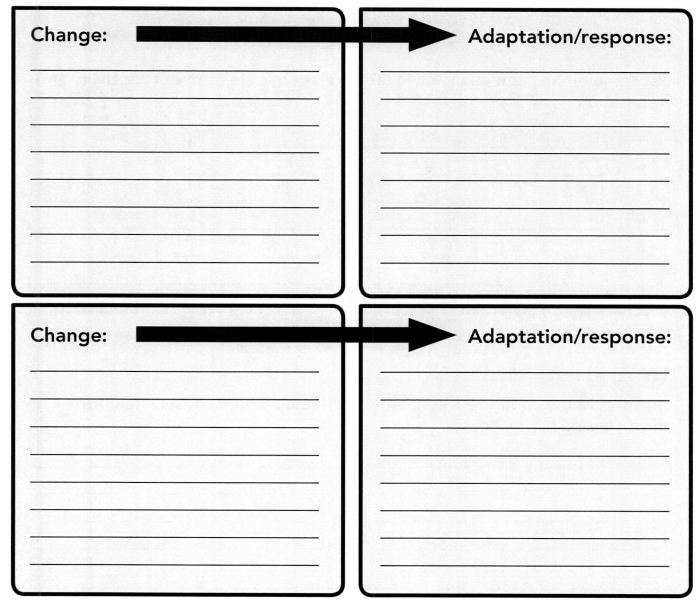

**Change:** ➡️ **Adaptation/response:**

**Change:** ➡️ **Adaptation/response:**

OTM2159   ISBN: 9781770783515
© On The Mark Press

Explain how changes in your aquatic life form's environment are monitored.

- who monitors the environment?
- what equipment do they use to monitor?
- how is this equipment used in the monitoring process?

_____

_____

_____

_____

_____

_____

_____

List some effects of human activity to your aquatic life form's productivity and distribution.

- _____

- _____

- _____

- _____

- _____

- _____

- _____

## Thinking critically...

How would the aquatic environment be affected if your chosen life form *disappeared*?

_____

_____

_____

_____

_____

OTM2159  ISBN: 9781770783515
© On The Mark Press